GUIDELINES FOR ACQUISITION AND MANAGEMENT OF BIOLOGICAL SPECIMENS

A Report of the
Participants of a
Conference on Voucher Specimen Management
sponsored under the auspices of
the Council on Curatorial Methods
of the
ASSOCIATION OF SYSTEMATICS COLLECTIONS

Edited by
Welton L. Lee
Bruce M. Bell
and
John F. Sutton

Published by the Association of Systematics Collections

ACQUISITION AND MANAGEMENT OF BIOLOGICAL SPECIMENS

QH
61
.C66
1981
c.2

GUIDELINES FOR ACQUISITION AND MANAGEMENT OF
BIOLOGICAL SPECIMENS
A Report of the Participants of a Conference on Voucher Specimen Management

Published May 3, 1982

asc

Copyright © by the Association of Systematics Collections
All Rights Reserved

No part of this book may be reproduced or transmitted in any form or by any electronic or mechanical means, including photocopy, xerography, recording, or by use of any information storage and retrieval system, without prior written permission of the publisher. The only exceptions are small sections that may be incorporated into book reviews.

ISBN 0-942924-02-9

Copies of **Guidelines For Acquisition and Management of Biologica Specimens**
can be ordered from:
  ASC
  Museum of Natural History
  University of Kansas
  Lawrence, Kansas 66045
  U.S.A.
  (913) 864-4867

PREFACE

Early in 1980, the ASC's Council on Curatorial Methods began to respond to a growing concern expressed in all quarters of the systematics community about the general lack of understanding of the need for quality voucher specimens to verify studies ranging from basic research to complex environmental studies. The failure to consider the importance of voucher specimens seriously compromises the reliability, accuracy, and ability to repeat otherwise good research. It is furthermore costing granting and contracting agencies valuable resource dollars in wasted and unusable information.

The Council responded to this problem by proposing to organize and convene an interdisciplinary conference to define these issues and develop guidelines for management of voucher specimens. The objective was to assemble a cross-section of the scientific, legal and administrative communities to assure broad consideration of all aspects of the problem. The project was funded by the National Science Foundation (Grant No. DEB-8020909) and following many months of research, evaluation and preparation, a three-day conference at the University of Maryland Conference Center was held September 23-25, 1981.

ACKNOWLEDGEMENTS

We wish to acknowledge the financial support provided by the National Science Foundation (Grant No. DEB-8020909) that made this workshop possible. In addition, we would like to recognize the extensive contributions of the participants in the workshop, both formally designated work group members and others who found it of value to contribute to this process. Each has contributed to the content and preparation of the report that follows.

Council on Curatorial Methods

Welton L. Lee, Chairman
California Academy of Sciences

Bruce M. Bell
Stovall Museum

John F. Sutton
St. Norbert College

Work Group Members

Sydney Anderson
American Museum of Natural History

Robert A. Bye, Jr.
University of Colorado

Bruce B. Collette
National Marine Fisheries Service
National Museum of Natural History
Smithsonian Institution
American Society of Ichthyologists
 and Herpetologists

K. Elaine Hoagland
Academy of Natural Sciences,
 Philadelphia
American Malacological Union

Lloyd Knutson
United States Department
 of Agriculture
Entomological Society of America

Alan Kohn
University of Washington
Society of Systematic Zoology

Micah Krichevsky
National Institute of Health
American Society for Microbiology

Wesley Lanyon
American Museum of Natural History
American Ornithologists' Union

Charles Huckins
Desert Botanic Garden
American Association of Botanical
 Gardens and Arboreta, Inc.

Samuel B. Jones
University of Georgia
American Society of Plant Taxonomists

David Kavanaugh
California Academy of Sciences
Coleopterists' Society
Entomological Society of America
Society of Systematic Zoology

K. C. Kim
Pennsylvania State University
Entomological Society of America

Paul M. Marsh
Agricultural Research Service
United States Department of Agriculture

C. J. McCoy
Carnegie Museum of Natural History
Herpetologists' League
Society for the Study of Amphibians
 and Reptiles

Frederic H. Nichols
United States Geological Survey

Mary Hanson Pritchard
Nebraska State Museum
American Society of Parasitologists

Michael Voorhies
Nebraska State Museum
Society of Vertebrate Paleontology

Terry Yates
University of New Mexico
American Society of Mammalogists

Other Participants

Daniel M. Cohen
National Marine Fisheries Service

Jack R. Coulson
Agricultural Research Service
United States Department
  of Agriculture

Stephen R. Edwards
Association of Systematics Collections

Alfred L. Gardner
Fish and Wildlife Service

C. G. Gruchy
National Museum of Natural Sciences,
  Ottawa

Steven R. Hill
University of Maryland

Meredith Jones
National Museum of Natural History
Smithsonian Institution

A. M. Neuner
Association of Systematics Collections

Paul Opler
United States Department
  of the Interior
Fish and Wildlife Service

Amy Y. Rossman
United States Department of Agriculture
Animal and Plant Health
  Inspection Service
National Fungus Collection

Robert Schoknecht
Cornell University

Graig D. Shaak
Florida State Museum

Greg Sohn
United States Geological Survey

D. W. S. Sutherland
United States Environmental Protection
  Agency

Robert Van Syoc
California Academy of Sciences

David L. Vincent
Agricultural Research Service
United States Department of Agriculture

Thomas E. Wallenmaier
Animal and Plant Health Inspection
  Service-Plant Protection and
  Quarantine
United States Department of Agriculture

TABLE OF CONTENTS

	Page
PREFACE	iii
ACKNOWLEDGEMENTS	v
TABLE OF CONTENTS	vii
INTRODUCTION	1
CLARIFICATION OF FEDERAL REGULATIONS	3
CHARACTERIZATION OF VOUCHER SPECIMENS	5
Definition and Importance	5
Categories of Voucher Specimens	5
Studies Requiring Voucher Specimens	6
Physical Forms of Voucher Specimens	6
Utility of Voucher Specimens	7
PLANNING	9
SCHOLARLY PUBLICATION POLICIES	11
PREPARATION OF VOUCHER SPECIMENS	13
Collection and Field Preservation	13
Field Notes	15
Data to Accompany Voucher Specimens	15
Taxonomic Identification	16
Specimen Maintenance	17
Preparation of Major Taxonomic Groups	17
SELECTION OF A REPOSITORY FOR VOUCHER SPECIMENS	19
Planning Ahead	19
Seeking Advice	20
Criteria for Repository Designation	20
Characteristics of a Suitable Repository	20
COSTS, FEES, AND FUNDING RESPONSIBILITIES	23
Cost Structure of Voucher Specimens	23
Cost Summary of Voucher Specimens	28
Costs of Managing Botanical Specimens as Voucher Collections	29
Fee Assessment and Schedules	31
SUMMARY	33
REFERENCES CITED	35
APPENDIX I	37
APPENDIX II	41
RECOMMENDATIONS	Inside back cover

INTRODUCTION

Biological and other natural history collections constitute a vital, non-renewable technological resource for the United States. Traditionally, these collections have been generated from taxonomic (classification of organisms) studies and other basic (non-applied) research projects. Today, these collections and the data associated with them are growing rapidly from new sources of specimens, *i.e.*, projects are directly funded by governmental agencies, others by the private sector, in fulfillment of recent federal and state legislation mandating an analysis of probable environmental impact prior to certain alterations of the environment. Many of these studies generate, or should generate, voucher specimens.

A voucher specimen is an organism or a sample thereof preserved to document data in an archival report. Unfortunately, little attention has been directed toward maximizing the potential of these materials or assessing the impact of their addition to collections in the nation's repositories. Many projects are conducted without a clear understanding of what a voucher specimen is, how it must be handled or preserved, what information must be assembled with it to make it of value, or what to do with vouchers after a project is completed. Vast numbers of specimens are being collected which hold the potential to greatly expand our resources and data base, thereby providing the American people with a clearer and more accurate analysis of our world and its problems.

The Council responded to this problem by proposing to organize and convene an interdisciplinary conference to define these issues and develop guidelines for management of voucher specimens. The objective was to assemble a cross-section of the scientific, legal and administrative communities to assure broad consideration of all aspects of the problem.

The assembled group (participants and their affiliations are listed in the Acknowledgements) combined the varied interests of the entire biological scientific community of the United States, and included representatives from the fields of paleontology, invertebrate zoology, biological control, entomology, parasitology, plant ecology, ornithology, malacology, microbiology, herpetology, plant taxonomy, mammalogy, and ichthyology. Also represented were major biological and systematic organizations, repository administrators and curators, governmental agencies that regulate or fund projects that may generate voucher specimens, as well as the following national societies:

American Association of Botanical Gardens and Arboreta
American Malacological Union
American Ornithologists' Union
American Society for Microbiology
American Society of Ichthyologists and Herpetologists
American Society of Mammalogists
American Society of Parasitologists
American Society of Plant Taxonomists
Association of Systematics Collections
Coleopterists' Society
Entomological Society of America
Herpetologists' League

North American Benthological Society
Society for the Study of Amphibians and Reptiles
Society of Systematic Zoology
Society of Vertebrate Paleontology

Although many participants were accustomed to working with preserved organisms, much of this report applies equally well to organisms kept alive for study and propagation.

The participants of the conference, organized into working groups, reviewed the many issues pertaining to voucher specimens, and produced by consensus the following report. The topics specifically addressed are:

* What legislation (Federal and State) governs the use of biological specimens?

* What constitutes a voucher specimen?

* What preliminary steps must be taken in projects involving environmental, ecological or other analyses that deal with organisms?

* What guidelines are available for proper collection and preservation of various kinds of specimens?

* What data should accompany collections of biological specimens?

* What are the characteristics of qualified institutions when selecting a facility to maintain biological specimens?

* What costs are involved in depositing and managing biological specimens?

A particularly important aspect of the report is the section pertaining to assessment of costs of depositing specimens in a repository and recommendations concerning who should pay these costs. Institutions housing systematics collections across the country have long been underfunded, although an acceptable, but far from optimal, level of care has been provided because of the dedication and pride of the research scientists managing these collections and the relatively modest rate of collection growth formerly observed. Recently, however, a dramatic acceleration of collection growth due to the vast numbers of voucher specimens now being collected has overwhelmed the physical and human resources represented by existing repositories. Valuable materials may be refused because funds for processing them are not available, or because inappropriately processed material may be too costly to reprocess Thus, deposition costs must be included among initial developmental costs of all projects to prevent loss of these vitally important voucher specimens.

It is the strong recommendation of the conference participants that these recommendations be incorporated as policy by all agencies that have as an objective the conduct of biological research (projects) requiring voucher specimens. Adoption would standardize procedures and distribute costs in an appropriate and equitable fashion. Failure to do so will result in the further loss of potentially important information, serious compromises of scientific accuracy and objectivity, and loss of money in unnecessary, ill-conceived projects.

CLARIFICATION OF FEDERAL REGULATIONS

There are no federal and very few state or local regulations specifically addressed to voucher collections as defined by this conference. Regulations reviewed included:

National Environmental Policy Act (NEPA), particularly as applied to environmental impact statements,

The Endangered Species Act,

National Historical Preservation Act,

National Antiquities Act, and

Regulations within the Department of Energy, Federal Energy Regulatory Commission, and the Fish and Wildlife Service as they relate to NEPA.

Such federal funding agencies as the National Institute of Health and the National Science Foundation were also contacted.

None of the federal regulations examined make any mention of voucher collections. Section 1502.24 of NEPA states that "... agencies shall insure the scientific integrity of environmental impact statements" but does not elaborate. The only statement even vaguely referring to collections is contained in Title 30 CFR-Wildlife and Fisheries, Section 17.22 concerning requirements for applications to make scientific studies on endangered species; the applicant must state "... the planned disposition of such wildlife upon termination of the activities sought to be authorized." The Migratory Bird Act makes a similar statement. The Sundry Civil Act for 1880 states "... all collections of rocks, minerals, soils, fossils, and objects of natural history, archaeology and ethnology, made by the Coast and Interior Survey, the Geological Survey, or by any other parties for the Government of the United States, when no longer needed for investigations in progress shall be deposited in the National Museum." Thus, although there are no specific federal guidelines concerning the need for making voucher collections, it is clear that such collections made by or for federal agencies must be offered for deposit in the National Museum of Natural History, Smithsonian Institution.

It is also apparent that there are very few state regulations concerning voucher specimens or specimen banking. Several agencies in Nebraska were unaware of the entire concept of voucher specimens and reported that they would defer to the federal government concerning voucher collections. This is probably true for many other states. California has no specific voucher specimen policy, although regional boards of the Water Resources Control Board may and do dictate their acquisition and maintenance on a case-by-case basis. The State of Texas, however, requires voucher collections of vertebrates for environmentally-related projects.

Some private companies involved in environmental assessment do appreciate the meaning of voucher specimens, but those contacted do not know of any state or federal laws governing the collection or maintenance of such specimens.

There is no direct federal and little state commitment to maintain or require voucher collections, although there is indirect commitment under the Sundry Civil Act of 1880. There is also indirect commitment in various federal and state requirements for obtaining permits to collect biological specimens where collectors are usually required to state the eventual deposition of specimens. One federal research laboratory -- The Beneficial Insect Introduction Laboratory, U.S. Department of Agriculture, Beltsville, Md. -- appreciates the need to provide a permanent record of the first release of any imported beneficial arthropod species into the U.S., and is now developing guidelines for establishing the U.S. National Voucher Collection of Introduced Beneficial Arthropods.

Although there are presently no regulations governing such activities, we believe that federal and state agencies **granting** research funds and awarding contracts for biological or environmental work where vouchers are necessary should require that documentation includes voucher material. Furthermore, they should require the collecting company, agency, or individual to follow a set of guidelines dealing with the proper methods of:

1. collection;
2. identification;
3. preparation, including required data elements; and,
4. selection of an appropriate repository.

These guidelines should apply not only to environmental studies, faunal surveys and ecological studies, but also to appropriate basic and applied research (e.g., behavioral and biochemical studies).

It should be the responsibility of agencies requiring or funding a study to see that the guidelines are followed. Costs covering acquisition and storage of voucher material should be included in the appropriate grants or contracts. A strong emphasis should be given to careful planning of such projects.

CHARACTERIZATION OF VOUCHER SPECIMENS

A voucher specimen is one which physically and permanently documents data in an archival report by:
1. verifying the identity of the organism(s) used in the study; and,
2. by so doing, ensures that a study which otherwise could not be repeated can be accurately reviewed or reassessed.

Kinds and numbers of voucher specimens adequate to document such a report should be determined jointly by participating investigator/collector(s) and repository institution(s) at the onset of the project.

Identification of organisms is the first step in communicating an investigator's results in any report involving any biological entities (Carriker, 1976; Hedgpeth, 1961; and Heppell, 1979). As noted in an earler report (Lee, et al., 1978):

"In all biological studies concern must be given to the accurate identification of the organisms under investigation, for this is the key to all past and future reference to data derived from the study. Given the often extreme variations in tolerances, biochemical and behavioral adaptations, and responses at all levels to normal and abnormal perturbations found between related but distinct species, genera, etc., accurate identification represents the only key to repeatability in the laboratory or the field. As such, it represents one of the foundations of the scientific approach and of its often extolled objectivity."

Voucher specimens ensure that identification of organisms studied can be verified and corrected if necessary even in cases where a study cannot or will not be repeated, such as before and after environmental studies, ecological studies, large faunal or floral surveys, or such studies as biochemical investigations where research involves destruction of the study specimen. They are the sole means to verify the data documented in a report and to make historical comparison possible. In addition, they provide critical information for future investigations of, for example, biochemical properties, demography, and geographic distribution. These properties make voucher collections unique resources that must be carefully protected.

The three categories of voucher specimens are:
1. type specimens, upon which names of taxonomic units are based;
2. taxonomic support specimens -- specimens of primary importance in taxonomic studies other than nomenclatural studies, such as range extensions, life-history studies and morphological variability; and,
3. biological documentation specimens -- representative organisms derived from studies or projects other than primarily taxonomic.

Voucher specimens are required in any study in which:
1. verification of experimental results can be accomplished only through reassessment or reevaluation of existing data, and where the species involved may not be unequivocally identified without access to samples taken in the original study;

2. the nature of the study brings about alteration of specimens such that future re-identification or verification is made impossible;
3. diversity of taxa under study is so great or the systematics so complex that all species involved probably will not be identified accurately;
4. taxonomic groups are not yet known accurately to species level; or,
5. the nature of the study is strictly systematics.

Studies requiring voucher specimens can be characterized as:

A. *Space/time specific studies*

Such studies determine organisms in a given location at a given time and include ecological base-line studies and environmental impact statements. The fact that populations fluctuate over generations, move, and become extinct are basic phenomena pertinent to such studies.

B. *Disciplinary studies*

These address phenomena characteristic of levels of biological organization ranging from molecular to ecosystem (*e.g.*, chemicals produced by particular cells, physiology of particular organ systems, behavior of particular organisms, and patterns of population
distribution and dynamics).

C. *Experimental studies*

These studies emphasize extraction, characterization, or manipulation of components or contents of organisms, or responses of organisms to stimuli. This is a special case where verification can **only** be achieved through vouchers. Without voucher specimens, such studies cannot be reproduced or critically evaluated, since the organisms utilized are destroyed through the experimental process or experimental organisms are returned to their environment.

D. *Systematic Studies*

These studies deal with problems of speciation, phylogeny and classification.

Voucher specimens may be:

1. **The actual organism** (part or whole) that is studied, observed or treated.
2. **A sample** of one or more individuals (part or whole) from a population that is studied, observed, or treated.
3. **A representation of the organism(s)** or its characters (*e.g.*, sound recordings, photographs, fossils, *etc.*) that is studied, observed, or treated. However, these representations are usually not adequate as a substitute for voucher specimens and should be used only when the organisms themselves are impractical or illegal to collect.
4. **An associated specimen** that is biologically or functionally related (*e.g.*, stomach contents, parasites, pollen preparations, *etc.*) to the organism that is studied, observed, or treated.

5. **A corroborative specimen** that provides additional data or character (*e.g.*, from the same population or individual but a different time or stage of the life cycle) to a previously collected voucher specimen (categories 1-4).

To fulfill its function, a voucher specimen must:

1. Have recognized diagnostic characters that are appropriate to the level of identification in the report. Specific life stages or body parts may be required.
2. Be preserved in good condition by the investigator/collector according to acceptable practice.
3. Be thoroughly documented with field and/or other relevant reports.
4. Be maintained in good condition and be readily accessible in suitable repository institution(s).

PLANNING

As with any scientific research, advance planning is essential to successful collection and storage of voucher specimens. With projects that generate voucher specimens, careful planning not only facilitates a smooth operation but does so at a lower cost than would be possible otherwise. Most importantly, such planning enhances accuracy and scientific merit of the project.

Planning should start with a thorough consideration of the reason for the project and its objectives. This is best done by framing a series of questions which can be answered through the activities of the project. As trivial as this may seem, projects with no clearly stated objectives lead to no clear answers and to useless collections. Many thousands of useless "vouchers" have been collected for projects designed to "look at" a situation or "survey" something.

The second step in planning is to decide whether or not vouchers are necessary. If they are, early contact should be made with a professional systematist, preferably at a potential repository.

Contact should be made with the repository(s) selected. Discussions should include the amount and nature of material to be deposited and costs involved. Furthermore, a series of decisions should be made regarding details of collection and preparation. These should include, but are not restricted to, basic data to be taken, collecting techniques to be used, choice of what is to become voucher material, nature of relaxation techniques, *etc.*, as well as form of preservation, level of field identification, and permits required. With such a review, projects can be streamlined to assure cost-effectiveness and scientific accuracy. Similarly, repository(s) can ensure that vouchers will be of the highest caliber, thus allowing for widest possible use by the scientific community as well as their important role of verifying project results.

Although the labor-intensive nature of survey work demands that lower cost personnel be utilized as much as possible, qualified and well-trained biologists are essential to planning a study and extremely valuable in collection and preparation of specimens. These steps must be carried out well to insure the continuing value of vouchers, and institutions may reject specimens if they are improperly collected or poorly processed.

SCHOLARLY PUBLICATION POLICIES

It was the consensus of the conference that recognition of the justification for and extreme importance of voucher specimens has been overlooked far too long. Accordingly, every effort should be made to incorporate these guidelines into the editorial practices of scientific publications. As noted earlier, when vouchers are necessary to document a project, yet are not collected and deposited in a suitable repository, scientific objectivity and accuracy are seriously compromised. All scientific societies responsible for scientific journals and/or publishers of said journals should:

1. acquaint themselves with the importance and relevance of vouchers;

2. publish as part of their instructions to authors the criteria used to decide whether vouchers are needed; and,

3. in those cases where vouchers are necessary, refuse publication unless the paper notes deposition of voucher material in a suitable repository.

PREPARATION OF VOUCHER SPECIMENS

Specimens are a vital part of the data base in any natural history investigation, easily as important as the written record. Yet, proper handling of specimens is often the most neglected part of the data gathering effort. Specimens become most critically important when, after the original investigation has been completed, new questions arise that are unrelated to the original questions and can be answered only by studying the original collection. Improper preparation and care of specimens means that such questions must go unanswered.

The task for those who collect specimens is to appreciate the importance of the specimen and to translate that appreciation into careful attention to techniques required to ensure permanence of specimens as a record.

Many years of experience have contributed to development of current standards for collection, storage, and preservation of specific taxonomic groups of organisms. Voucher specimens encompass all biological taxa, and techniques for collection, handling and preservation vary widely from one taxonomic group to another; thus it is not feasible to formulate specific standards for voucher specimens in general. But use of existing standards for each group of organisms must be required. In addition, holding institutions, over the years, have developed their own standards for housing, use, and loan of specimens that are in their care, and these standards must be followed.

We discuss here general approaches to collecting and preparing voucher specimens and associated supporting data, and direct the prospective collector to special sources of information and procedures unique to major taxonomic groups.

Collection and Field Preservation

Collection and initial processing of a specimen is critical to its later usefulness. Great care must be taken to insure that the collection techniques:

1. are not damaging to individual specimens collected (even slight damage could render specimens useless as a permanent research resource);
2. do not unknowingly bias the results obtained (for example, census techniques that selectively sample some species or size groups and unintentionally overlook others); and,
3. do not involve destructive sampling when rare or fragile habitats are studied.

If longterm storage and use are to be guaranteed, application of correct procedures is highly critical. This is accomplished best through work done under the direction of trained biologists. A trained biologist is one who knows or is aware of proper field procedures and techniques associated with the biological taxa involved, is aware of federal and state regulations pertaining to the taking of organisms from the wild, including the requirement of collecting and import permits, and is intimately familiar with techniques required to meet requirements for eventual longterm disposition of specimens. Examples of the lack of training are all too common (Finley, 1980).

Careful attention must be given to requirements of the institution housing the specimens. Most institutions will accept only material that has been processed by approved techniques, and moreover, will refuse specimens that have not been collected, preserved, or identified according to standards established for the taxa involved. Therefore, a suitable repository should be found before specimens are collected. Representatives of that institution must then approve plans for processing this material and accept responsibility for its long-term maintenance. They should decide on the voucher material to be deposited as this will vary depending upon the taxa involved and the nature of the study.

Some material is best returned to the laboratory or museum alive (nematodes, slime molds, bacteria or other microorganisms) while other material (e.g., zooplankton) should be preserved immediately upon collection.

Assuming that most live material cannot be carried to the laboratory for processing without damage, the initial processing of a specimen must be done in the field. Preparation and handling of field samples may include one or more of the following processes, depending on the taxa being collected and the envisioned use of the material:

1. **Relaxation** of (living) animal specimens to prevent distortion of whole animals and/or individual tissues or organs, prior to fixation of specimens.

2. **Isolation** of microorganisms or other live organisms for propagation and future culture.

3. **Initial fixation** of specimens requiring fixation in the field, appropriate to the type of final disposition. Future use of specimens will dictate the type of fixation (and final preservation). For example, analysis for trace metals or other biochemical analyses will require specific preservation techniques that will not alter or contaminate the material. Often, freezing or freeze-drying is the method of choice rather than use of organic preservatives. Fixation is not necessarily equivalent to preservation; different chemicals may be required for the two processes. It is vital to preserve recognized diagnostic characteristics that are appropriate to the level of identification necessary.

4. **Packaging** of specimens that allows for safe transport to a laboratory for appropriate processing. Care should be taken to eliminate undesirable foreign material (e.g., insect pests) from specimens before packaging. Special care must be taken for pathogens or toxic material. Contact postal authorities for proper methods of shipping such materials.

5. **Complete preservation** of specimens in the field in the manner used for final storage may be required.

6. **Final laboratory disposition,** which may include the following:

    a. whole specimen or section slide mounts,

    b. whole specimen or tissue storage in fluid,

    c. whole specimen storage as dry material,

    d. whole specimen or tissue storage for chemical component analysis.

7. **Photography** is often a valuable additional means of documenting specimens, especially for rare taxa or those that do not preserve well (*e.g.*, nudibranch mollusks; orchids). However, photography is usually not a substitute for taking voucher specimens. There are standards for taking photographs of some taxa, such as birds.*

Field Notes

For a specimen to be of use in a permanent collection, it must be accompanied by a complete set of notes characterizing the site from which it was taken and the conditions under which it was collected. Such information can be very extensive, depending on the objectives of the initial investigation. The following information is pertinent to preparation of voucher specimens:

1. A record must be permanently but nondestructively associated with the specimen, lot, or sample. This data record must be unequivocally related to the specimens. Specimen label and ink must be made of material that will withstand any chemical preservative to be used.
2. The identity and concentrations of chemicals used in fixation and preservation processes must be part of the data record, contained on the label affixed to the specimens, if possible.

Data to Accompany Voucher Specimens

Based on work of the ASC Council on Standards for Systematics Collections (Anon., 1975), we have formulated the following set of guidelines for mandatory categories of data which must accompany biological specimens to make them useful to investigators other than the collector. These specimen data are basic to, but exclusive of, documents such as scientific papers, summary reports, impact statements, *etc.*, which summarize purpose, findings and conclusions of the original project.

1. **A unique sample designation** must be assigned during the collection process to each sample collected at one place and time. This designation will be supplemented by later repository designations (see category 8).
2. **The position of a sample collection site** shall be described so that this site can be revisited. Recommended elements of locality description are:
    a. Continent or ocean;
    b. Country or oceanic region;
    c. State, province, county, or other political subdivisions;
    d. Sea, major island group, river system, geologic horizon;
    e. Local place names;
    f. Latitude and longitude or Universal Transverse Mercator grid designation;
    g. Altitude or depth; and,
    h. Habitat including (where applicable) host, bioassociations, substratum, *etc.*

---

*Dr. Frank Gill of the Academy of Natural Sciences (Philadelphia) can provide information on Project Vireo, a computerized system of bird photographs used as voucher material.

3. **Time and date of sample collection** as well as any other biologically significant dates such as date of preservation, propagation, isolation, *etc.*

4. **Name of collector** (*e.g.*, person, vessel, project, expedition) and other donor identification including station or field numbers.

5. In general, and where feasible, **identity to species level** and name of person who made the identification and the date (year) of identification should be recorded. The demands of some environmental impact statements could be met by identifications at the generic or higher taxonomic level. The level of identification possible will also be dictated by scientific expertise available at the time.

6. **Method of collection and preparation.** This must include specific descriptions of equipment, techniques, preservatives, fixatives, isolation media, *etc.*

7. If codes are necessary, consider using a **standard coding system** and **always** maintain copies of translations, definitions of all codes, station numbers, *etc.* Unequivocal information must accompany samples that allow translation and/or definition of any and all codes or mnemonics used in recording data.

8. Every **repository** must have a distinct identifier, both for itself and the specific item (*e.g.*, ANSP 123456; USNM, Fish Division, Tank 7). This identifier is assigned by the repository in addition to, and irrespective of, the collector's identifier.

Other kinds of data, such as color notes, ecological characteristics and metabolic features may be useful to subsequent investigators and should be considered for inclusion in the data accompanying specimens, although these data are not as essential as categories 1-8 listed above.

The original label with basic data outlined above must accompany a voucher specimen at all times so it can be accurately associated with the documentary material it validates. Two-way tracing, from specimen to report and from report to specimen is essential. We suggest that a federally supported central data bank be established. All collectors of specimens for impact statements should register their collections to make these data accessible to researchers. This central clearing house would publish and circulate, possibly in a scientific journal, a notice that such biological collections were being made, and where and when the specimens would be available.

Taxonomic Identification

Any company, individual, or agency anticipating a study involving collection of voucher specimens must obtain advice from an appropriate qualified taxonomist(s) before the study to establish the level of identification needed and to determine the expertise available for the taxa involved. The level of identification needed will be determined by the purpose of the study. Specific identification of an organism could be critical, as in the case of an outbreak of a pest or disease agent. On the other hand, the requirement of some environmental impact statements could be met by identifications at the generic or a higher level. The level of identification possible will also be dictated by the scientific expertise available at the time. Furthermore, the collecting company or individual must also be required to finance the taxonomic services provided for the study and **plan the budget** accordingly.

Specimen Maintenance

The voucher specimen lot, together with its number and associated field notes, comprises a one-of-a-kind scientific record that requires storage and maintenance for an indefinite period while providing a unique resource for:

1. continuing study and review of conditions existent at the time of collection;
2. verification or change of the originally-designated taxonomic name; or,
3. new investigations of the taxa (species or higher taxa).

The specimen lot, label, and field notes may be later supplemented by written reports, but **all original labels must be retained.** Curatorial maintenance may include:

1. ensuring appropriateness of the initial preservation and altering it when necessary;
2. routine restoration of preservative fluids;
3. fumigation; * and,
4. providing additional labels which conform to institutional standards and cataloging material for incorporation into the institution's collections.

This work represents a long-term financial obligation for the institution. Therefore, the decision to incorporate voucher material into an existing collection must remain with the institution.

Preparation of Major Taxonomic Groups

Appendix I lists some recommended available manuals describing the initial preparation for long-term storage of specimens of major taxonomic groups. It is our view that familiarity with these manuals should be required as part of the training of, and used in the field by, field collectors and research scientists involved in collecting voucher material. Use of these manuals will not guarantee proper preparation and care in all cases, but will increase collectors' awareness that initial collection and processing of specimens is critical to their longterm usefulness.

---

*Consult Edwards, S. R., et al. (1980)

SELECTION OF A REPOSITORY FOR VOUCHER SPECIMENS

A suitable repository is an institutional collection serving two basic functions:

1. preserving specimens and related information; and,
2. making it easy for people to find and use them later. A good repository does these things well; a poor one does not.

Planning Ahead

The person, group or agency seeking a future repository for voucher specimens should carefully consider the following questions as early as possible **before** collection begins:

1. What will be the nature and quantity of material and data collected?
2. Where is the material likely to be of greatest future use?
3. Where are the biological specialists who might help in the present study or use the materials in the future?
4. Are identifications expected? If so, to what level?
5. What will be the scheduling and terms of the deposit of such materials?
6. How should specimens be processed prior to deposition?
7. Are there to be any restrictions on future use?
8. Can the specimens be integrated into the general collection or must they be kept separately?
9. Are supporting funds available for costs of processing, future care, or other services?
10. Will the institution you have in mind as a repository accept your collection? Remember that although the initial selection of a repository is the investigator's/collector's responsibility, final decision on acceptance of material must remain with the institution. It is not obligated to take all materials that might be sent.

Although federal regulations (Title 30 CFR-Wildlife and Fisheries: Section 17.22) state that applications for scientific studies must declare "... the planned disposition of such wildlife upon termination of the activities sought to be authorized," there is no statement on appropriate repositories. The only specific comment on repository designation is in the Sundry Civil Act of 1880 which states that organisms collected by or for the United States government shall be deposited in the United States National Museum. This 100 year old act of Congress is not as relevant today as it was when the scientific community was concentrated in the northeastern part of the country. Voucher material often has regional significance, and federally supported regional collection centers for collections mandated by the federal government (environmental impact statements, endangered and threatened species studies and importation of foreign species) would make voucher collections more accessible and more useful to investigators.

Seeking Advice

With these general considerations or requirements in mind, contact an appropriate professional organization for guidance. Unless a specific relevant disciplinary group is already known, consult the Association of Systematics Collections (ASC, Museum of Natural History, University of Kansas, Lawrence, KS 66045; 913 864-4867). One or more suitable repositories will be found among its member institutions. Sometimes a single repository is all that is needed. This may be the case when only one type of material (such as preserved fishes) is involved. In most cases, however, surveys are broadly ecological and preparations are diverse. Thus, several different institutions may need to be involved as repositories. In these cases, it is difficult to keep information linked unless there is careful planning.

The *Encyclopedia of Associations* (Gale Research Company, Book Tower, Detroit, MI 48226) available in most reference libraries is another source of information. Most professional societies in biology are listed there.

There are published lists of museums that provide useful background information also. For example, see *The Directory of World Museums* (Columbia University Press, 1975).

The American Institute of Biological Sciences (AIBS) (1401 Wilson Blvd., Arlington, VA 22209) is an association of biological societies. It is a good contact for advice on sources of biological services other than for collection-related questions that can be answered more easily by the ASC. The AIBS, for example, contracts with government agencies for consultation on biological questions, recruitment of advisory panels of experts for special studies, conduct of surveys, and other services.

Criteria for Repository Designation

Professional persons, institutions, disciplinary groups, umbrella organizations, or other reference sources will recommend or suggest possible repositories. Actual selection of a repository should be made by the depositor after discussion of detailed terms of the deposit with curators of the potential recipient institution.

In selecting a repository, a potential depositor should be aware of and should apply the criteria listed below rather than rely upon uncritical advice. The reason for this is that people who planned the study that produced the voucher material conducted the field work, prepared the specimens, and used the newly gained information in preparing reports, understand the ramifications of their information system better than anyone else. These ramifications should be discussed with the relevant curator at potential recipient institutions. These curators may have ideas for optimizing future values of the material, ideas that have not occurred to the originators, hence the desirability of such collaboration.

Characteristics of a Suitable Repository

A. The collection is administered by a non-profit public or private institution.

B. The collection has at least one curator who is directly responsible for it.

ACQUISITION AND MANAGEMENT OF BIOLOGICAL SPECIMENS

C. The collection is housed in a building that provides adequate protection from fire, water, dust, excessive heat or light, and other physical hazards. Important permanent records (such as catalogs and field notes) should be kept in a fireproof or fire retardant safe or its equivalent.

D. Specimens are stored in containers appropriate to the discipline and nature of the specimen.

E. Specimens are periodically inspected and maintained in accordance with accepted procedures.

F. Specimens are prepared and processed in a manner that insures their present and future utility.

G. Specimens are arranged according to a specific plan that is recorded and, preferably, posted.

H. Field notes and ancillary data are preserved as a part of the permanent record for each specimen or lot.

I. Data on specimen labels, in field notes, in permanent catalogs, and wherever else data are recorded in the collection are accurate and original labels are available to investigators.

J. The collection is accessible to all qualified users. Qualified users are those with demonstrated ability to handle material properly, and a specific purpose for so doing.

K. Accessibility to collections by unqualified persons is restricted. We recommend formation of separate teaching collections for use in basic courses, and restriction of other materials to research use.

L. Loans to other institutions are handled and packaged in an appropriate and legal manner. When material is kept alive and propagation is possible, the repository agrees to provide subcultures, progeny, or seeds to qualified researchers.

M. Type specimens are identified as such, segregated and marked accordingly, and made accessible only to qualified scientists. Loans, if made, are considered and handled with special care.

N. The institution has the stated intent to continue support of the collection at least at a level necessary to maintain these standards. Should institutional priorities change, the institution will transfer the collection to an appropriate institution which will insure its perpetual maintenance.

O. Specimens are acquired and possessed in accord with federal and state regulations pertaining thereto.

P. A written policy exists for collection management, including acquisition, preservation, and de-accessioning.

The above criteria ensure that the vouchers stored will be well maintained and remain accessible to future investigators. For this reason, established repositories that meet these criteria should receive all voucher collections; new repositories should not be constructed for the sole purpose of voucher collection maintenance. Appendix II lists references helpful in the selection of an appropriate repository.

COSTS, FEES, AND FUNDING RESPONSIBILITIES

Specimen storage and maintenance are complex tasks and very expensive.. Voucher specimens acquired through biological, environmental impact or archeological inventory studies may well become extremely numerous and difficult for many institutions to store and manage properly. Accordingly, voucher specimens should be of high scientific quality and their storage and maintenance will have to be highly selective. To accomplish this, qualified taxonomists or representatives of the voucher repository should be involved with the project from the outset.

The cost of making voucher specimens available is a composite of a series of expenses incurred from the collection to final deposition and maintenance. These costs may vary greatly depending on the nature of the study and vouchers taken, amount of planning prior to outset of the study, size and organizational structure of the repository.

Conference participants unanimously agreed that fees should be charged to accept, process and maintain voucher collections and these costs should be borne by the contracting or funding agencies. However, once voucher specimens are properly prepared, deposited, and integrated into the main collection of a repository, they are of value to the housing institution and to future investigators, and thus overhead and indirect costs should be the responsibility of the repository.

The following assessment of fees is based on opinions gathered (by questionnaire) from curators, collection managers, and directors of natural history museums, herbaria, university collections, and botanical gardens, systematic biologists whose research is based on systematics collections, environmental consultants and state officials. Since such costs are known to vary, their figures represent only guidelines. Furthermore, these figures will inevitably increase due to inflation.

Cost Structure of Voucher Specimens

There are generally five major categories of the costs for voucher specimens:

A. Collection and Preparation;

B. Accessioning, Processing and Identification;

C. Maintenance and Management; and,

D. Use, Propagation, and/or Application.

The following sections will deal with each of these categories.

A. Collection and Preparation

These costs depend largely upon how specimens are collected and prepared by the collector. To reduce costs and ensure high quality specimens, a representative of a qualified repository should be consulted to assist in the proposal formulation and research or survey design. The contractor should be prepared to pay for this service.

Actual costs for collection and preparation of voucher specimens vary considerably depending upon objectives and magnitude of different projects, type and condition of specimens, and level of taxonomic service required. Some of these factors have been discussed comprehensively for various programs, for example, Leupke (1979) on environmental monitoring.

If specimens are properly preserved and labeled during the collection process and are to be deposited without further processing in a collection center, initial costs for accessioning and processing voucher specimens should be reasonably low. However, if field samples are improperly collected and preserved by lots or bulk, processing costs will be considerably higher.

All projects should require that specimens be properly prepared for prompt taxonomic use, and costs for material and labor should be included in the cost structure of voucher specimens.

Prior to submitting a project proposal which involves collection of voucher specimens, arrangements should be made with a repository institution for accessioning of these vouchers. We recommend that these arrangements be verified prior to funding.

B. **Accession, Processing and Identification**

Upon arrival at a collection center or museum, voucher specimens must undergo several routine preparatory steps whose sequence may vary for different institutions. These steps are identified in Figure 1 and follow the model for preparation of mammals developed by Anderson (1973). The procedure is a labor-intensive task, requiring primarily support personnel except for step 7 which requires a curator or qualified taxonomist.

Accession includes those processes associated with the first three steps of Figure 1, from unpacking specimens after receipt to recording them in accession books or a master catalog. Specimens are examined and mode of deposition is determined after the shipment is received and unpacked. The process further involves acknowledging receipt of specimens and taking and recording appropriate data.

The cost estimate for processing a specimen, which includes preparing labels, placing specimens in temporary storage for fumigation, preparation and eventual incorporation into the collection (steps 4, 5) will also vary greatly according to types and conditions of specimens, and the way they were collected (Stuessy and Thomson 1981). Processing a fox specimen will certainly be many times more expensive than processing a grasshopper. Large collections which do not require specific identification will cost much less per specimen than vouchers requiring in-depth taxonomic service. Step 5 involves preparing specimens for permanent deposition into the collection. Specimens are properly preserved (dried or placed in a final preservative), mounted (tanned, pinned or affixed), and labeled. Cost for preparation includes material and labor expenses.

Examples of items associated with preparation costs are:
1. Fixative and preservative -- Ethyl alcohol (75%), formalin (10%), and fumigants.
2. Materials and supplies -- jars and vials, herbarium sheets, pins, microscope slides and cover glasses, unit trays, labels, mounting media, wires, chemicals (stains, dehydrating agents, tanning agents, *etc.*), papers, and others.
3. Equipment -- drawers, shelving, cabinets, microscopes, *etc.*

4. Labor --
   a. loading, unpacking;
   b. sorting and recording;
   c. preparing (mounting, pinning, skinning, labeling, *etc.*);
   d. inserting or incorporating; and,
   e. administration.

Computer related expenses, when definitely a part of the preparation process (*e.g.*, cataloging) in an institution, must be considered as a legitimate processing expense.

FIGURE 1. Processes involved in accession and processing of voucher specimens.

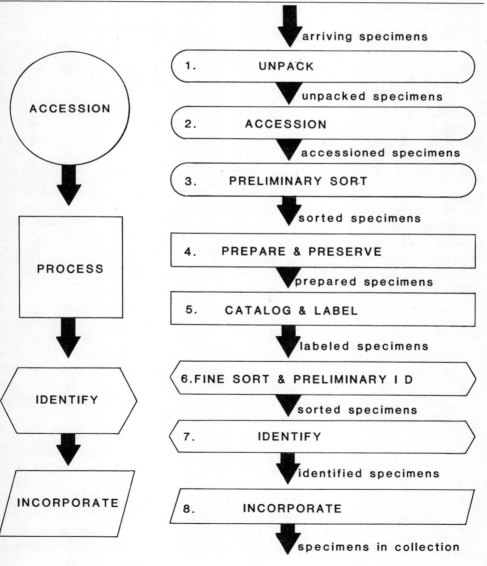

Considering all these factors, the following estimates for accession and preparation cost per specimen (excluding identification fees) are cited for reference (all figures based on 1981 costs):

1. Birds and mammals .... $15-$20
2. Plants .... $2-$2.50
3. Insects:
   a. pinned .... $1.20-$1.50
   b. slide .... $2.50-$4.50
   c. vial with preservative .... $1.50-$2.50
4. Invertebrates other than insects:
   a. wet-preserved .... $2.50-$5.25
   b. dry .... $1.15-$2
5. Microorganisms .... $25-$50 per subculture (depending on level of external subsidy)

Costs of identification services (step 7) may vary significantly (Chilson 1978). Common species can be identified without much effort by a specialist or even by a general taxonomist and this service may cost as little as $10-$20 per hour, whereas species which require a literature search and detailed examination could cost as much as $20-$100 per hour, dependent upon type and condition of the specimen.

Final incorporation of voucher specimens into the permanent main collection involves retrieving specimens from temporary voucher storage, sorting and arranging according to the filing system, and finally putting into cabinets or shelves. This step primarily involves labor costs; for example, two years ago incorporation cost was estimated by the Advisory Committee for Systematics Resources in Botany, American Society of Plant Taxonomists (ACSRB/ASPT 1979) to be $0.17-$0.20 for a plant specimen.

All cost figures given above will inevitably increase due to inflation.

C. Maintenance and Management

Once voucher specimens are received, they can be integrated into the main collection with specific voucher identifiers to allow for easy retrieval. However, when a voucher collection needs to be kept separate from the main collection for an extended period of time, it may be necessary to assess special fees for maintenance and management.

Normal collection management involves inspecting the condition of specimens, repairing and mending, identifying and rearranging specimens because of taxonomic revision or new additions; propagation of microorganisms for subcultures, fixing specimens with liquid nitrogen or freeze-drying, adding more preservative and pest deterrent agents or fumigating, and preparing exchanges; the input and updating of data storage, maintaining, and administering programs, personnel and research. In those cases where acceptance of voucher material forces the repository to incur additional expenses for expansion (additional space needs, storage equipment, *etc.*), it is reasonable to supplement the capital investments necessary through special grants from federal and state agencies.

One of the costs listed above is fumigation, which refers to two different operations, one involved with destroying pests on all incoming or infested material, and the other with keeping specimens and storage cases free from infestation. Unrecognized pests associated with incoming material should be sent to specialists for identification. Pests new to a state or region should be sent to the Animal and Plant Health Inspection Service, Plant Protection and Quarantine, U.S.D.A., Federal Center Building, Hyattsville, MD. All incoming or infested specimens must be taken to a fumigator or treated with fumigant by the staff. Two years ago, fumigation costs for plant material were estimated to be $0.73 per thousand sheets (ACSRB/ASPT 1979). The maintenance operation typically involves preparing pest deterrent (such as "No-Pest Strips"), and inserting it into the cases (Edwards *et al.*, 1980). There are no comparable figures on maintenance fumigation; however, information so far available suggests costs for maintenance operation may be very high.

To some extent, costs for routine management service will also depend upon collection management practices (*e.g.*, separate voucher collections), quality of technical manpower, curatorial standards and procedures, and administrative overhead of a museum or collection center. These costs could be calculated on the basis of curatorial units, for example, per specimen, drawer or cabinet. Furthermore, actual costs should be negotiated prior to proposal formulation between contractor and repository institution.

D. Use or Application of Voucher Specimens

Voucher specimens may be used in several different ways: a specialist or user may wish to visit the museum or collection center to study voucher specimens, borrow specimens of a specific taxonomic group for study, obtain bacteria or tissue cultures, use specimens for chemical, electrophoretic or other analyses, or obtain information on particular species. Costs for these uses are normally borne by the host organization if no profit to the user is involved. If the examiner will reap some financial benefit from his study of the vouchers, he should expect to reimburse the host organization for expenses resulting from the visit. Costs should also include any additional expenses incurred in dissemination of data associated with voucher collections to the scientific and user community. Thus, costs for use or application of voucher specimens will differ according to institutional size, curatorial practices, and scientific activity generated by the collection.

One of the most important services of voucher collection centers is that of lending specimens to workers who cannot visit the institution in person. The process of preparing outgoing loans and receiving returning loans is fairly complex and expensive in labor cost. The size of a loan affects the cost. Outgoing loan cost has been estimated (ACSRB/ASPT, 1979) as follows for plant specimens:

```
1 specimen loan ........................................$4.68
2-25 specimen loan ..................additional $0.68 per specimen
26-250 specimen loan ...............additional $0.123 per specimen
```

When loaned specimens are returned, three categories of action are carried out before specimens are reinstated into the main collection:

1. specimen handling;
2. record keeping; and,
3. preparation for reinsertion, including fumigation.

Processing cost for a returned loan was estimated to be $5.71 per loan for plant specimens and the insertion cost was 17.5 cents per specimen (ACSRB/ASPT 1979). Visitor accommodation costs in herbaria were estimated to be $17.58 per visitor (based on 1978 costs) for larger herbaria, and $12.28 for small herbaria.

Cost Summary of Voucher Specimens

The real cost of an average voucher specimen may be calculated on the basis of parameters considered in the preceding pages:

General costs per specimen = collection * + accessioning + preparation * m processing m identification * + incorporation + overhead (maintenance, management, and use).

Using this formula, general total cost for a voucher specimen can be estimated (figures based on 1981 costs):

Birds and mammals	$43
Plants	$5.50
Insects:	
pinned	$3.50
slide	$9
vial	$5
Invertebrates other than insects:	
wet-preserved	$5-$11
dry	$4
Bacteria Cultures	$25-50 per subculture (depending on the level of external subsidy)

As an example of the distribution of costs incurred in managing voucher collections, a recent comprehensive analysis of the costs of managing botanical voucher specimens made by the ASCRB/ASPT is included here. Annual costs have been estimated for processing and housing newly collected plant specimens from tropical regions (Steussy and Thomson, 1981).

---

* These parameters may be borne by the contractor or subcontracted to qualified individuals.

Costs of Managing Botanical Specimens as Voucher Collections

The most accurate, credible and expedient means of determining costs of voucher collections involving botanical specimens would be to apply the costs of herbarium services developed by the ACSRB/ASPT in its 1979 report published by the New York Botanical Garden, Cary Arboretum, Millbrook, New York, entitled, "Systematic Botany Resources in America, Part II, the Costs of Services." These must be updated by using federal information figures. Such costs are divided into four major categories:

1. specimen actions for routine service needs (including accessioning, mounting, insertion of materials, outgoing loans, returned loans, outgoing exchange, incoming exchange and fumigation);
2. information services (including requested identification services, routine identification and visitor accommodation);
3. purchase of storage equipment; and,
4. maintenance of physical plant.

Costs for maintaining botanical collections representing national resources have been found to be most accurately and easily determined on a per specimen basis. A summary of 1979 costs for maintaining botanical collections, primarily on a per specimen basis, follows. These figures do not include expenditures relating to collection of specimens; rather, they relate to cost of preparing specimens for permanent storage, retrieval and use as research tools. The degree to which these expenses might justifiably be applied to the cost of managing voucher collections would probably depend most upon the nature of a particular agreement between a contracting agency and a holding agency.

A. Specimen actions for routine service needs:

per specimen cost

	Average	Range
1. Accessioning		
Large herbaria (1-4.5 million specimens)	$00.09	$00.04 - $00.11
Smaller herbaria (1 million specimens or fewer)	$ .19	.02 - .77
All herbaria	.14	.02 - .77
2. Mounting		
Large herbaria	.83	.46 - 1.81
Smaller herbaria	.79	.23 - 1.86
All herbaria	.81	.23 - 1.86
3. Insertion		
Large herbaria	.17	.06 - .38

Smaller herbaria	.20	.09 - .42
All herbaria	.19	.06 - .42

4. Outgoing loans

Bulk loans (1000 + sheets)	.13	.02 - .15
Individual loans	4.86	2.08 - 7.45

5. Returned loans

Costs without insertion (Per loan)	5.71
Insertion costs (per sheet)	.18
Total cost returned loan (one specimen)	5.89

6. Outgoing exchange

Large herbaria	.33
Smaller herbaria	.37
All herbaria	.35

7. Incoming exchange (per thousand specimens) 73

B. Information services

1. Requested identification services (cost per ID)

Large herbaria	3.57
Smaller herbaria	4.59

2. Routine identification (cost per ID)

Large herbaria	1.40
Smaller herbaria	2.38

3. Visitor accomodation (per visitor)

Large herbaria	17.58
Smaller herbaria	12.28

C. Purchase of storage equipment

(per additional specimen) 20

D. Maintenance of physical plant . . . . 50% of costs for all other categories

Fee Assessment and Schedules

Actual costs of voucher specimens will vary greatly between taxonomic groups and from institution to institution as shown in the preceding section. A prior establishment of fee assessment and schedules for all organisms would not only be a major undertaking but would change so frequently due to inflation and geographic locality that it would be impractical. Actual costs are best considered on a case by case basis, and should be negotiated prior to proposal formulation. Under certain conditions institutions may charge a consultation fee for assistance in proposal preparation.

In most cases, costs of preparation, identification, accession, cataloging, and processing (including any required equipment, materials or labor) should be paid for by project funds.

Costs of long-term maintenance, management, use and application are usually absorbed by the holding institution, except when said use and application are for profit. In this case, a user or consulting fee may be charged at the discretion of the holding institution. Likewise, when voucher collections need to be maintained separately from the main collection (at the contractor's request) for an extended period, any additional costs should be borne by the project. Once incorporated into the main collection, the housing institution should be responsible for maintenance costs since it will gain the greatest long-term benefits.

No justification exists for random accumulation of enormous numbers of specimens in national and regional repositories. As shown in the foregoing pages, costs for collection, processing and maintaining voucher specimens are considerable. However, the costs of ignoring the importance of representative samples (voucher specimens) can be substantial, both monetarily and in terms of information loss.

SUMMARY

The scientific integrity of a research publication depends on the ability of subsequent investigators to repeat the study described. Thus, the identification of the organism(s) studied is the first step in communicating the results of a biological investigation. By preserving examples of the organisms involved, the investigator creates "vouchers" or specimens which permit verification or duplication of a project or study and provide critical information for future scientific investigations.

There appears to be little governmental commitment to maintain or to require voucher collections. Nevertheless, if the capacity to verify and to repeat or critically evaluate environmental and other projects (including basic and applied research) is desired, then voucher specimens must be taken and recognized standards of collection, preservation and storage must be observed. Many years of experience have contributed to development of current standards for specific taxonomic groups of organisms. This report summarizes those standards and includes an annotated bibliography for techniques used in preparing museum specimens for each major group of organisms.

Planning for collection, preparation, storage and allocation of costs of vouchers is as important as the planning of experimental design or survey methodology. Voucher specimens should be collected by, or under supervision of, trained biologists. Specimens must be processed according to established standards for the taxa involved, and deposited in an appropriate repository.

Selection of a repository for voucher specimens involves finding an institutional collection that is suitable because it will preserve specimens and related information, and hold them in trust for subsequent users. Eighteen criteria are suggested to measure the adequacy of a repository. Arrangements with a repository (or repositories) should be made as early as possible in the planning stage of a project.

No repository is obligated to accept specimens, especially those that have not been collected and prepared according to standards established for the taxa. Consultation with a qualified taxonomist from the repository during the project planning phase will establish specific requirements for preparation and handling of voucher material as well as levels of identification needed.

Voucher material must be accompanied by a permanent label or other data record that unequivocally relates the material to associated information and includes eight minimum data categories.

In addition, it is recommended that a central data bank be established where voucher collections relating to environmental impact and assessment statements would be registered.

Six major categories of costs for voucher specimens are presented. Guidelines are presented, but actual costs of developing and maintaining voucher collections can be determined only on a case by case basis after consultation between contractor and repository. All financial arrangements for professional consultations, as well as handling and long term maintenance of material should be made prior to initiation of a project.

Expenses can be reduced greatly and quality of the project vastly improved with proper planning before a collection program begins. The above costs should be written into all project proposals and these costs should be honored by funding agencies. Likewise, collection and deposition of vouchers, where appropriate, should be required by funding agencies.

A list of recommendations concerning voucher specimens was compiled and is included on the inside back cover.

REFERENCES CITED

Advisory Committee for Systematic Resources in Botany, Amer. Soc. Plant Taxonomists (ACSRB/ASPT). 1979. Systematic botany resources in America. Part II. The cost of services. A report to the public, and to the botanical systematics community. The New York Botanical Garden, Cary Arboretum, Millbrook, New York. i-vii, 116 pp.

Anderson, S. 1973. It costs more to store a whale than a mouse: libraries, collections and the cost of knowledge. Curator 16: 30-44.

Anonymous. 1975. Report of the ASC Council on Standards for Systematics Collections *in* ASC Newsletter 3(3).

Carriker, M. R. 1976. The crucial role of systematics in assessing pollution effects on the biological utilization of estuaries. Pages 487-506 *in* (U.S.E.P.A., Office of Water Planning and Standards) Estuarine pollution control and assessment, Proceedings of conference, Volumes 1 and 2, U.S. Government Printing Office, Washington, D.C.

Chilson, L. M. 1978. Report of the ESA Standing Committee on Systematics Resources in entomology: Fee charges for identification. Bulletin of the Entomological Society of America 24:167-169.

Directory of World Museums. 1975. Columbia University Press, New York. xviii + 864 pp.

Edwards, S. R., B. M. Bell, and M. E. King, eds. 1981. Pest control in museums: a status report (1980). The Association of Systematics Collections. Lawrence, Kansas. 34 pp.   7 appendices.

Encyclopedia of Associations. 1974. 9th ed. Gale Research Company, Detroit.

Finley, R. B. Jr. 1980. Deficiencies and training needs in nongame wildlife management. Pages 268-270 *in* K. Sabol, ed. Transactions of the 45th North American Wildlife and Natural Resources Conference. Wildlife Management Institute, Washington, D.C. 478 pp.

Hedgpeth, J. W. 1961. Taxonomy: Man's oldest profession. Eleventh annual University of the Pacific Faculty Research Lecture, May 22, 1961.

Heppell, R. 1979. Biological collections, systematics and taxonomy. Museum Journal 79(2):75-77.

Lee, W. L., D. M. Devaney, W. K. Emerson, V. R. Ferris, C. W. Hart, Jr., E. N. Kozloff, F. H. Nichols, D. L. Pawson, D. F. Soule, and R. M. Woollacott. 1978. Resources in invertebrate systematics. American Zoologist 18(1): 167-185.

Luepke, N. P. ed. 1979. Monitoring environmental materials and specimen banking. Proceedings of the International Workshop, Berlin (West), 23-28 October 1978. Marinus Nijhoff, Hague, Netherlands.

Stuessy, T. F., and K. S. Thomson. 1981. Trends, priorities, and needs in systematic biology. A report to the Systematic Biology Program of the National Science Foundation. Association of Systematics Collections, Lawrence, Kansas. 51 pp.

APPENDIX I
ANNOTATED BIBLIOGRAPHY TO SELECTED REFERENCES
Preparation of Specimens in Major Taxonomic Groups

A. Mammals:

Hall, E. R. 1962. Collecting and preparing study specimens of vertebrates. University of Kansas Museum of Natural History, Miscellaneous Publications 30. 46 pp. (Pages 13-25 cover mammals.)

Preparation of study skins is the traditional method described here. Photography may be an alternate method when technical assistance and/or storage facilities are unavailable. Standards for photography would have to be established to insure that photographs are comparable to one another and to study-skins.

B. Birds:

Hall, E. R. 1962. Collecting and preparing study specimens of vertebrates. University of Kansas Museum of Natural History, Miscellaneous Publications 30. 46 pp. (Pages 26-35 cover birds.)

This reference deals primarily with preparation of study-skins.

C. Reptiles and Amphibians:

Hall, E. R. 1962. Collecting and preparing study specimens of vertebrates. University of Kansas Museum of Natural History, Miscellaneous Publications 30. 46 pp. (Pages 37-40 cover reptiles and amphibians.)

Liquid preservatives are apparently still the method of choice. They can be used in the field.

Pisani, George R. 1973. A guide to preservation techniques for amphibians and reptiles. Society for the Study of Amphibians and Reptiles, Herpetological Circulars No. 1. 22 pp.

A guide to the preservation techniques for amphibians and reptiles.

D. Fishes:

Hall, E. R. 1962. Collecting and preparing study specimens of vertebrates. University of Kansas Museum of Natural History, Miscellaneous Publications 30. 46 pp. (Pages 42-44 cover fishes.)

Fink, W. L., K. E. Hartel, W. G. Saul, E. M. Moon, and E. O. Wiley. unpubl. A report on current supplies and practices used in curation of ichthyological collections. American Society of Ichthyologists and Herpetologists, Ichthyological Collection Committee, *ad hoc* Subcommittee on Curatorial Supplies and Practices. 63 pp.

A report on current supplies and practices used in curation of ichthyological collections.

E. Aquatic and Marine Invertebrates:

Chivers, D. D. (chairman), Robert Setzer, Jack Ward, Eric Hochberg, and Robert Lavenberg. Unpubl. (1976) Guidelines for marine ecological surveys. California Committee on Marine Ecological Survey Standards. California Sea Grant, Marine Advisory Programs, University of California. Davis, California. (Pages 16-18 cover invertebrates.)

Taxa with calcareous structures must be stored in buffered preservative. Some taxa become unusable when stored in alcohol. Each phylum is handled differently. Relaxation prior to fixation is critical for soft-bodied invertebrates, including mollusks.

Lincoln, Roger J., and J. Gordon Sheals. 1979. Invertebrate animals, collection and preservation. British Museum (Natural History), London and The Syndics of the Cambridge University Press, Cambridge. 150 pp.

Handbook of collecting methods and fixing and preservation techniques. Discussion of various preservatives, labelling and packing for shipment for invertebrates from Protozoa to Urochordata.

F. Mollusks:

Solem, A., W. K. Emerson, B. Roth, and F. G. Thompson, 1981. Standards for Malacological Collections. Curator 24(1):19-28.

Preservation must include fixation and storage of body as well as shell material.

G. Insects:

Borror, D. J., D. M. DeLong, and C. A. Triplehorn. 1981. An Introduction to the Study of Insects. 5th ed. Saunders College Publishing Company, Philadelphia. 827 pp. (Pages 711-753 cover preparation techniques.)

In most cases, dry storage is used, but preservation in alcohol is appropriate for some groups and most immature stages.

Martin, J. E. H. 1977. Collecting, preparing, and preserving insects, mites and spiders.

Collecting, preparing, and preserving insects, mites and spiders.

H. Nematodes:

Thome, G. 1961. Principles of Nematology. McGraw-Hill, New York. 553 pp.

Southey, J. R., ed. 1970. Laboratory methods for work with plant and soil nematodes. Great Britain, Ministry of Agriculture, Fisheries, and Food. Technical Bulletin no. 2:1-282.

Laboratory methods for work with plant and soil nematodes.

I. Parasites:

Pritchard, M. H., and G. O. W. Kruse. (1982) in press. Collection and preservation of animal parasites. University of Nebraska Press.

This is the only recent paper dedicated to the subject. Collecting shipping, staining, and mounting are among the topics covered.

J. Vascular Plants:

Jones, S. B. Jr., and A. E. Luchsinger. 1979. Plant systematics. McGraw-Hill, New York. xi 388 pp.

Chapter eight covers basics of preparation in the field and storage in the herbarium. Long-term storage requires use of insecticides. Jones and Luchsinger provide several references at the end of the chapter, including those by Archer, Bailey, Franks, Savile, and Smith.

Fosberg, F. R., and M. H. Sachet. 1965. Manual for tropical herbaria. Regnum Vegetabile 39:29-50.

This work treats special problems involved in preservation of plants in tropical environments.

K. Bryophytes:

Conrad, H. S. and P. L. Redfearn, Jr. 1979. How to know the mosses and liverworts. 2nd ed. W. C. Brown Co., Dubuque, Iowa.

Grout, A. J. 1972. Mosses with hand-lens and microscope. E. Lundberg, August, West Virginia. 416 pp.

Both references describe use of envelopes to contain dried specimens. The second reference mentions that a poison should be added to any glue used to prevent its being attacked by insects.

L. Algae:

Chivers, D. D. (chairman), Robert Setzer, Jack Ward, Eric Hochberg, and Robert Lavenberg. Unpubl. (1976) Guidelines for marine ecological surveys. California Committee on Marine Ecological Survey Standards. California Sea Grant, Marine Advisory Programs, University of California. Davis, California. (Pages 12-16 cover algae.)

Macroscopic algae can be dried and pressed as with vascular plants, freeze-dried, or kept in weak formalin. Alcohol destroys pigments.

M. Fungi:

Hanlin, R. T. 1972. Preservation of fungi by freeze-drying. Bulletin of the Torrey Botanical Club. 99(1):23-27.

>Air-drying is the traditional method; Hanlin cites several advantages of freeze-drying, including better retention of natural characteristics. Freeze-drying techniques can certainly be applied to other botanical material if the apparatus is available.

Dade, H. A. 1960. Laboratory methods in use in the culture collection, C.M.I. Pages 40-69 in Herbarium I.M.I. Handbook. Commonwealth Agriculture Bureaux, Somm-Wherry Press, Norwich, England. (Pages 40-69 cover fungi.)

>This reference describes methods for maintaining living fungi for taxonomic reference purposes, and for preparing slides.

N. Microorganisms:

American Phytopathological Society. 1981. Report of the national work conference on culture collections of importance to agriculture. St. Paul. 52 pp.

>Summarizes and compares preservation techniques of microorganisms, including viruses and cell cultures. In addition to this report, a pamphlet is available from the American Type Culture Collection that describes shipping of cultures of microorganisms. Other information should be solicited from the American Type Culture Collection.

O. General:

Knudsen, J. W. 1972. Collecting and preserving plants and animals. Harper and Row, New York. 320 pp.

>A good general reference covering field techniques, collection, relaxing, fixing, and preserving a wide range of taxonomic groups.

Steedman, H. F., ed. 1976. Monographs on oceanographic methodology. 4. Zooplankton fixation and preservation. UNESCO Press, Paris. 350 pp.

>Specific organisms discussed are zooplankters, but chapters on narcotizing agents, freeze-drying and aldehydes may have a much broader application. Bibliographies at the end of each chapter.

Hower, Rolland O. 1979. Freeze-drying biological specimens: a laboratory manual. Smithsonian Institution Press. Washington, D.C. 196 pp.

>Technical manual of freeze-drying principles, apparatus and techniques for various taxa.

APPENDIX II
PUBLISHED REFERENCES FOR SELECTING A REPOSITORY

A. Mammals:

Choate, Jerry R. 1978. Revised minimal standards, and the systematic collections that meet them. Journal of Mammalogy 59(4):911-914.

Choate, Jerry R. and Hugh H. Genoways. 1975. Collections of recent mammals in North America. Journal of Mammalogy 56(2):452-502.

B. Birds:

Banks, R. C., M. H. Clench, and J. C. Barlow. 1973. Bird collections in the United States and Canada. The Auk 90(1):136-170.

Clench, Mary H., Richard C. Banks, and Jon C. Barlow. 1976. Bird collections in the United States and Canada: addenda and corrigenda. The Auk 93(1):126-129.

King, James R., and Walter J. Bock. 1978. Workshop on a national plan for ornithology: final report, submitted to the National Science Foundation and the Council of the American Ornithologist's Union. 300 pp.

C. Reptiles:

Committee on Resources in Herpetology. 1975. Collections of preserved amphibians and reptiles in the United States. Society for the Study of Amphibians and Reptiles, Miscellaneous Publications, Herpetological Circulars No. 3. 22 pp.

Wake, David B., R. G. Zweifel, H. C. Dessauer, G. W. Nace, E. R. Pianka, G. B. Rabb, R. Ruibal, J. W. Wright, and G. R. Zug. 1975. Report of the Committee on Resources in Herpetology. Copeia 1975(2):391-404.

D. Fishes:

Collette, Bruce B. and Ernest A. Lachner. 1976. Fish collections in the United States and Canada. Copeia 1975(3):625-642.

Lachner, Ernest A., James W. Atz, George W. Barlow, Bruce B. Collette, Robert J. Lavenberg, C. Richard Robins, and R. Jack Schultz. 1976. A national plan for ichthyology (abstracted from a report to the Advisory Committee of the Development of a National Plan for Ichthyology). Copeia 1976(3):618-625.

E. Invertebrates:

Emerson, W. K., and A. Ross. 1965. Invertebrate collections: Trash or treasure? Curator 8:333-346.

Lee, W. L., D. M. Devaney, W. K. Emerson, V. R. Ferris, C. W. Hart, Jr., E. N. Kozloff, F. H. Nichols, D. L. Pawson, D. F. Soule, and R. M. Woollacott. 1978. Resources in invertebrate systematics. American Zoologist 18(1): 167-185.

Randolph, D. E., compiler. 1974. North American arachnid collections. Committee on Systematic Collections, American Arachnological Society. 16 pp.

Ward, R. A., O. S. Flint, Jr., A. S. Menke, and F. C. Thompson. 1976. The United States National Entomological Collections. Smithsonian Institution Press, Washington, D.C. 47 pp.

F. Fossil Animals:

Paleontological Society, *ad hoc* Committee on North American Resources in Invertebrate Paleontology (CONARIP). 1977. Fossil invertebrates -- collections in North American repositories, 1976. University of Iowa, Iowa City. 67 pp.

Society of Vertebrate Paleontology, *ad hoc* Committee on the Status of Fossil Vertebrate Conservation in the United States. 1972. Fossil vertebrates in the United States, 1972. 154 pp.

Society of Vertebrate Paleontology. 1977. Systematic resources in vertebrate paleontology. Fossil vertebrates in the United States: the next ten years. 40 pp.

G. Plants:

Holmgren, Patricia K., and Wil Keuken, compilers. 1974. Index Herbariorum, a guide to the location and contents of the world's public herbaria. Part I: The herbaria of the world, 6th ed. Oosthoek, Scheltema and Holkema, Utrecht, Netherlands. 397 pp.

Advisory Committee for Systematic Resources in Botany, Amer. Soc. Plant Taxonomists (ACSRB/ASPT). 1979. Systematic botany resources in America. Part II. The cost of services. A report to the public, and to the botanical systematics community. The New York Botanical Garden, Cary Arboretum, Millbrook, New York. i-vii, 116 pp.

H. General:

Conference of Directors of Systematic Collections. 1971. The systematic biology collections of the United States: an essential resource. Part I. The great collections: their nature, importance, condition and future. A report to the National Science Foundation. xi + 33 pp.

Cross, W. H., D. R. Dewey, P. A. Fryxell, A. M. Golden, R. E. Hanneman, Jr., G. B. Hewitt, P. L. Lentz, J. R. Lichtenfels, F. G. Meyer, D. R. Miller, F. D. Parker, T. G. Pridham, and C. R. Gunn (chair). 1977. Systematic collections of the Agricultural Research Service. U.S. Department of Agriculture, Agricultural Research Service, Miscellaneous Publications No. 1343. 84 pp.